Lab Manual – Universal Fluid Power Trainer

Modeling and Simulation for Application Engineers

Lab Manual
Universal Fluid Power Trainer (UFPT)
Modeling and Simulation for Application Engineers

ISBN: 978-0-9977634-5-4

Printed in the United States of America

Lab Manual – Universal Fluid Power Trainer

Hydraulic Systems Modeling and Simulation for Application Engineers

Lab Manual-UFPT

How to Use Me Safely?

1- Emergency Stop:
- Locate the emergency stop.
- Press the emergency stop immediately if you feel any dangerous situation e.g. leakage, bad smell, system instability, unusual noise

2- Warning Light: consult the instructor if any of red led lights is turned ON.

3- Consult the Instructor: if you are not sure about any thing.

Please, no use of flash drives

Safety Regulations, Contd.

MSOE – Fluid Power and Motion Control Professional Education

4. Safety Glasses: Wear the safety glass and shoes where applicable.

5. Hydraulic Hoses: Use **proper length &** respect minimum **bend radius.**

6. Quick Disconnect: Make sure it is perfectly connected **"hear the click".**

7. Trapped pressure: Connect high pressure measurement hose to the tank first, then to the point at which the pressure is trapped.

8. Accumulator (stored energy hazard): Isolate the accumulator whenever it is not needed.

9. Pressure Measurement Gauges: P4 assigned for tank pressure only using female-female hose between the tank header and P4.

10. Drain: make sure drain lines are plumbed where needed.

11. Flow-meter: is a unidirectional flow meter, follow the arrow.

12. Turning ON any power : Warn all team members first.

13. Circuit (Hydraulic/pneumatic/24V) Modification: turn the corresponding power button Off first.

Safety Regulations, Contd. For courses other than Introduction

14. Cables & Sockets: Keep the unused electrical sockets covered.

15. Cables & Sockets: Do not stretch or pull from wires & make sure it is correctly fit.

16. Servo Valve Cable: Servo valve cable contains DC/DC convertor.

17. Banana Jacks Cables: Use proper length of cables.

18. Electrical 24V Circuits: In order not to get confused and to avoid a short circuit, the good advice is to connect one vertical line at a time.

19. Pressure Sensors: make sure it is tightened enough to sense the pressure. High pressure sensors are numbered (1/4 – 3/4) and one low pressure is tagged (1/4).

20. Pneumatic Power: turn air valves off before disassemble air hose.

21. Compressor(stored energy hazard): Before turning the compressor ON, make sure air valves are off.

22. Follow up the "lab procedure".

23. Follow the machine "Machine Startup Procedure".

24. Follow the machine "Machine Shutdown Procedure".

Lab Procedure

1. Form work groups. A group works together till the end of the class..
2. Listen to the instructor orientation. Ask questions if needed.
3. Read the lab instructions line by line even after the orientation.
4. Read schematics (Electrical/Hydraulics) and work instructions.
5. Build the system as per the step-by-step given instructions.
6. Operate the circuits and follow the given instructions.
7. Recorded your observations as per the exercise instructions.
8. Analyze the data and answer the posted questions.
9. Disconnect the circuit and put down hoses and measuring instruments at the end of every lab session.
10. Store back the components as per its drawer identification.

Lab Procedure, Contd.

Hydraulic Power Supply

Adjustment of PRV at the beginning of every exercise

Matlab RT Models
1-Activating, 2-Connecting 3-Running

Matlab

Current Folder: C:\11-Common Matlab\Models Real Time

Simulink ▶ 100 External

Machine Startup Procedure

1. Casters: Lock the casters to prevent accidental move of the unit.
2. Uncoil the power cords and plug the unit to the electrical wall outlet.
3. Take the hoses out of the hose bin and hang them on the side hooks.
4. Turn ON the electrical control panel.
5. Turn ON the HMI (only when needed).
6. Turn ON the printer (only when needed).

6

Machine Shutdown Procedure

1. Make sure cylinders are fully retracted.
2. Discharge the accumulator from the manual discharge valve.
3. Discharge the compressor by opening one of the air valves.
4. Turn OFF the pump, air power, and control power.
5. Shutdown the computer from the windows.
6. Turn OFF the mouse and store it back in drawer "A"
7. Turn OFF the 24 Volt power supply.
8. Turn OFF the printer.
9. Set toggle switches to 'Pot" position.
10. Set the potentiometers to "min".
11. Unplug the power cord and hang it under the counter.
12. Unplug any remaining electrical connections.

7

Machine Shutdown Procedure, Contd.

Exit

13. Disconnect hydraulic hoses, cover their ends by dust caps, and bring the store hoses back to the hose bin.
14. Cover all hydraulic components with dust caps.
15. Store the components as per its drawer identification.
16. Please, clean the safety glasses and store them.
17. Please, clean spilled oil.

8

MSOE – Fluid Power and Motion Control Professional Education

Lab09: Pump Static Characteristic Measuring

Objective:
This lab is to practice measuring the static characteristics of a variable displacement pump.

Step 1: Prepare Components:

MSOE#-X	Component Name	QTY	Drawer #
N027	Pressure measurement hose	1	B
N016-TX-2/4	Flow meter (Medium Scale)	1	D2
N036	Proportional FCV + Cable	1	D5
N053	BNC Cable	1	D1

0

Step 2: Adjust the PRV to 500 psi:

1. Connect the pressure measurement hose as shown to P1
2. Set the Pp and Qp toggle switches to "Pot" position
3. Adjust pump potentiometers to Pp = max and Qp = max
4. Fully open the PRV
5. Turn ON the hydraulic power
6. Close the PRV gradually until the pressure gauge reads the set value.
7. Turn OFF the hydraulic power

1

Step 3: Electrical Circuit:

AI14 V1

1. Connect the proportional flow control valve cable to socket AI14V1 and the other side of the cable to the valve.
2. Put the toggle switch of V1 at "Ext." position
3. Connect the BNC Cable as shown below

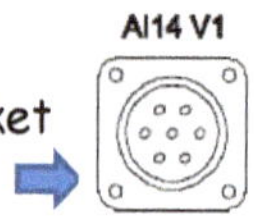

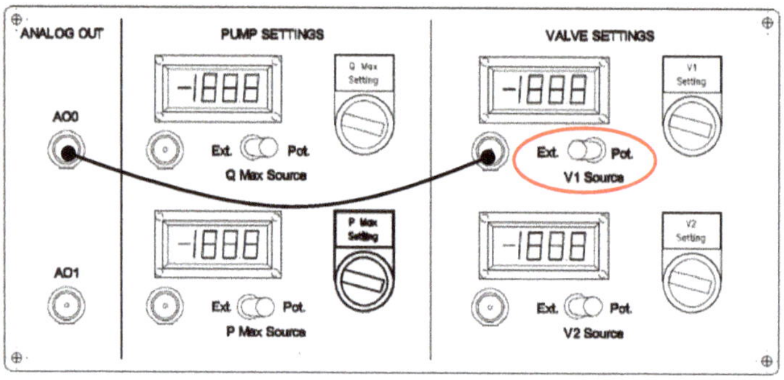

2

Step 4: Hydraulic Circuit:

1- Build the shown hydraulic circuit.
- Note: PFCV is 2-way mode, port p is disconnected
2- Turn the hydraulic power ON.

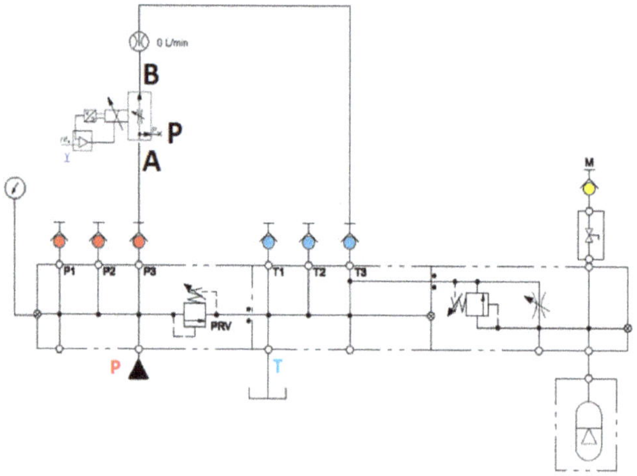

3

Step 5: Pump Static Characteristics Data Acquisition:

1. Activate,: RTModel011.mdl

2. Connect and run the model

3. Model will automatically plot the pump flow against variable pressure and stop after 100 s.

4. Turn OFF the hydraulic power

5. Use the plotted data to fill in the following table and draw the characteristic curve.

Note1: This characteristics is based on the current adjustments of Pmax. and Qmax.

Note2: The plotted flow value represents the pump total flow which is branched between the PRV and the Proportional FCV.

4

Step 6: Use the plot to fill in the pump static characteristics lookup table:

P (psi)	0	100	150	200	250	300	350	400	450	500
Q (L/min)										

5

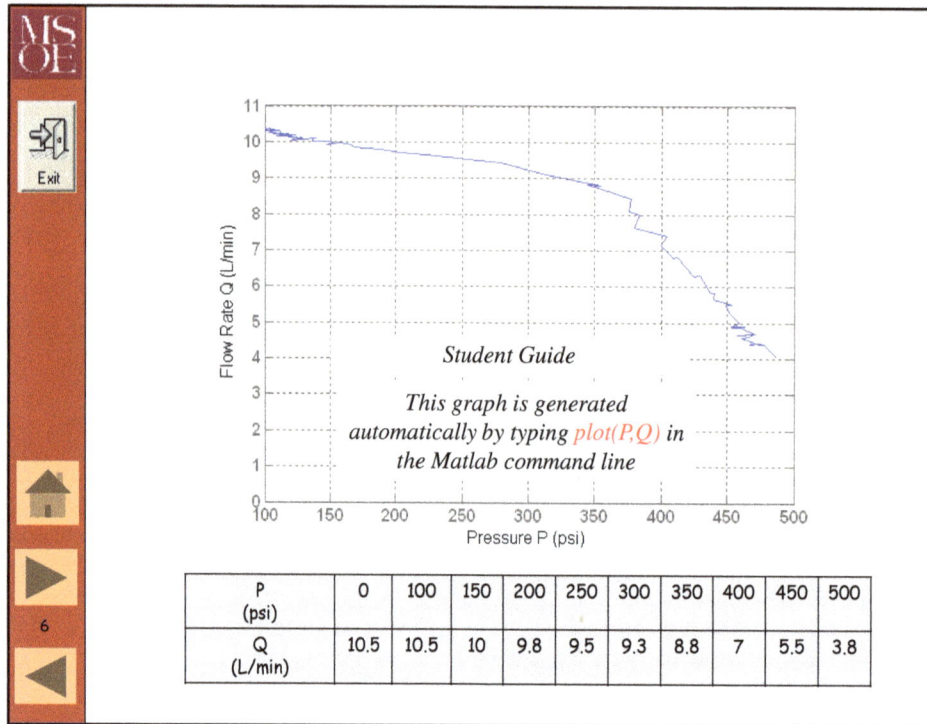

P (psi)	0	100	150	200	250	300	350	400	450	500
Q (L/min)	10.5	10.5	10	9.8	9.5	9.3	8.8	7	5.5	3.8

6

Lab10: Pump Step Response Measuring

Objective:
This lab is to practice measuring the step response of a variable displacement pump.

Step 1: Prepare Components:

MSOE#-X	Component Name	QTY	Drawer #
N027	Pressure measurement hose	1	B
N016-TX-2/4	Flowmeter (Medium Scale)	1	D2
N036	Proportional FCV + Cable	1	D5
N053	BNC Cable	2	D1

Step 2: Adjust the PRV to 500 psi:

1. Connect the pressure measurement hose as show to P1
2. Set the Pp and Qp toggle switches to "Pot" position
3. Adjust pump potentiometers to Pp = max and Qp = max
4. Fully open the PRV
5. Turn ON the hydraulic power
6. Close the PRV gradually until the pressure gauge reads the set value.
7. Turn OFF the hydraulic power

Step 3: Electrical Circuit:

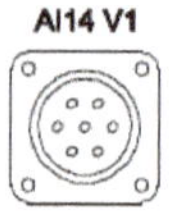

AI14 V1

1. Connect the proportional flow control valve cable to socket AI14V1 and the other side of the cable to the valve.
2. Put the toggle switch of Qmax and Pmax at "Pot." position.
3. Make sure Qmax and Pmax potentiometer is at max (10 Volt).
4. Put the toggle switch of V1 at "Ext." position.
5. Connect the BNC Cable as shown below.

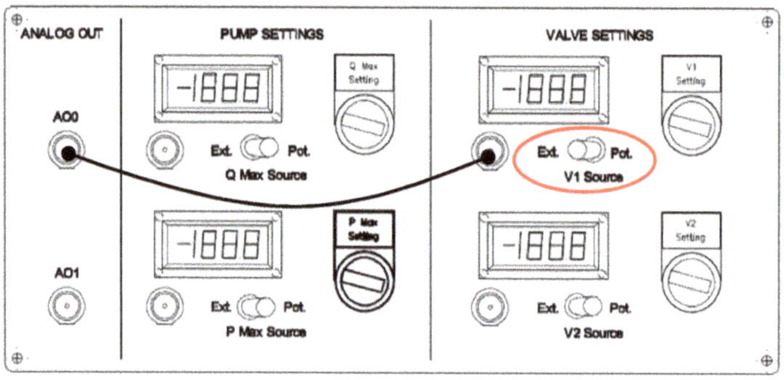

Step 4: Hydraulic Circuit:

1- Build the shown hydraulic circuit.

• Note: PFCV is 2-way mode, port p is disconnected

2- Turn the hydraulic power ON.

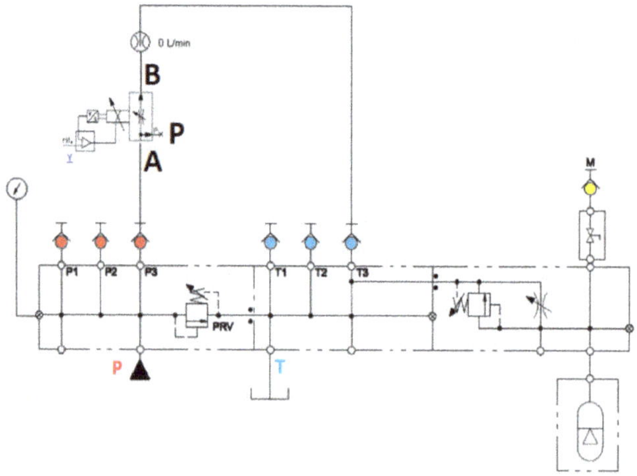

Step 5: Data acquisition. Pump Step Response

1. Activate,: RTModel016.mdl
2. Connect and run the model
3. Model will automatically plot the pump flow against variable pressure and stop after 40 s.
4. Turn OFF the hydraulic power
5. Observe the P-Q relation of the pump and share your observation with the instructor.
6. Type in Matlab command line plot(T,Q/11)

Lab11: Hydraulic Motor I-n Static Characteristics

Objective:

This lab is to practice developing the I-n static characteristics of a bidirectional fixed displacement motor.

Step 1: Prepare Components:

MSOE#-X	Component Name	QTY	Drawer #
N027	Pressure measurement hose	1	B
N033	Servo Valve + Cable with Interface Box	1	D5

Exit

0

0

Step 2: Adjust the PRV to 500 psi:

1. Connect the pressure measurement hose as shown to P1
2. Set the Pp and Qp toggle switches to "Pot" position
3. Adjust pump potentiometers to Pp = max and Qp = max
4. Fully open the PRV
5. Turn ON the hydraulic power
6. Close the PRV gradually until the pressure gauge reads the set value.
7. Turn OFF the hydraulic power

Exit

1

1

Step 3: Electrical Circuit:

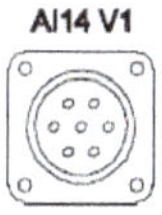

AI14 V1

1. Connect the servo valve cable to socket AI14V1 and the other side of the cable to the valve.
2. Put the toggle switch of Qmax and Pmax at "Pot." position
3. Make sure Qmax and Pmax potentiometer is at max (10 Volt)
4. Put the toggle switch of V1 at "Pot." position

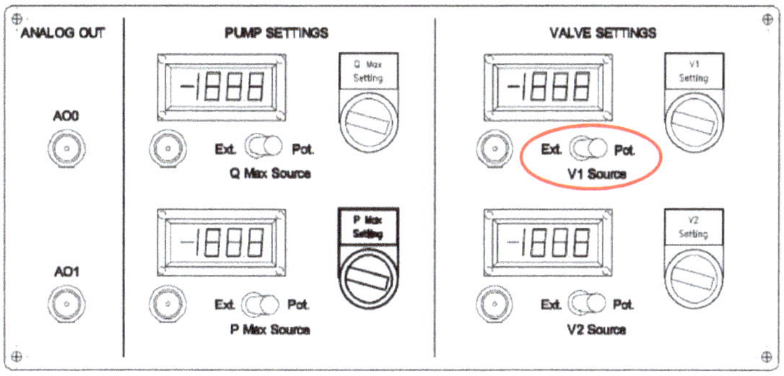

2

Step 4: Hydraulic Circuit:

1- Build the shown hydraulic circuit.

2- Turn the hydraulic power ON.

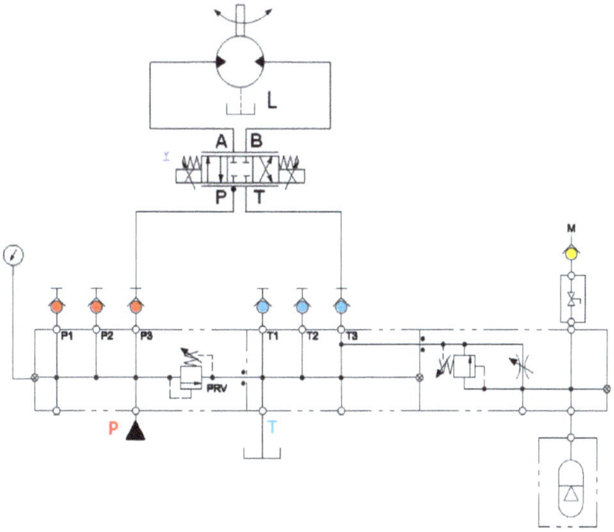

3

Step 5: Motor Static Characteristics Data Acquisition:

1. Chang the valve input signal manually using the potentiometer to meet the values shown in the table & record the corresponding RPM.
2. Use the table to draw the curve. Note: 2 quads for bidirectional motor

I (Volt)	RPM
0	
1	
2	
3	
4	
5	
6	
7	
8	
9	
10	

4

Guidance to Students

I (Volt)	RPM
0	0
1	0
2	165
3	265
4	347
5	428
6	516
7	590
8	662
9	715
10	765

5

Lab12- Identify Hydraulic Motor Dynamics

Objective:

This lab is to practice identification of motor dynamics

Step 1: Prepare Components:

MSOE#-X	Component Name	QTY	Drawer #
N027	Pressure measurement hose	1	B
N033	Servo Valve + Servo Valve Cable	1	D5
N053	BNC Cable	1	D1

0

Step 2: Adjust the PRV to 500 psi:

1. Connect the pressure measurement hose as shown to P1
2. Set the Pp and Qp toggle switches to "Pot" position
3. Adjust pump potentiometers to Pp = max and Qp = max
4. Fully open the PRV
5. Turn ON the hydraulic power
6. Close the PRV gradually until the pressure gauge reads the set value.
7. Turn OFF the hydraulic power

1

Step 3: Electrical Circuit:

1. Connect the servo valve to the socket AI14V1 and the other side of the cable to the valve.
2. Connect the BNC cable as shown below
3. Put the toggle switch of V1 at "Ext." position

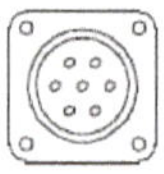

AI14 V1

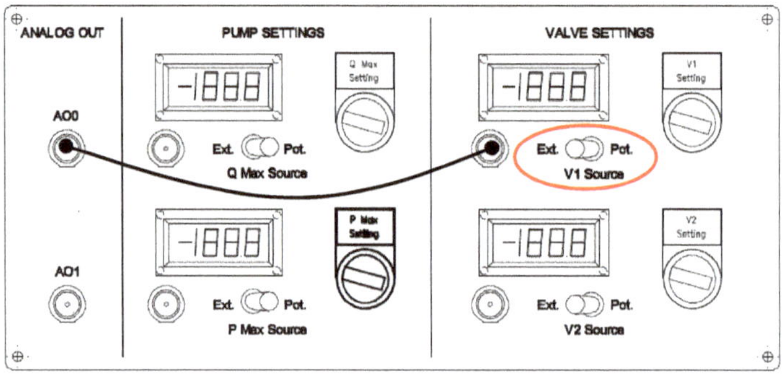

2

Step 4: Hydraulic Circuit:

1- Build the shown hydraulic circuit.
2- Turn the hydraulic power ON.

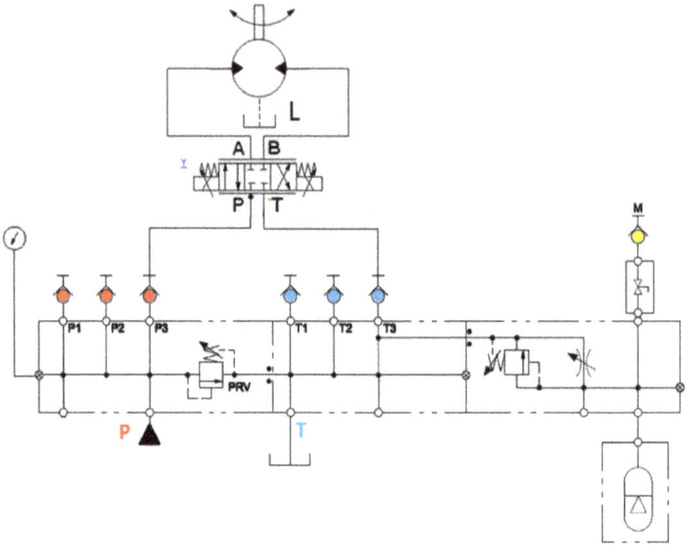

3

Step 5: Data Acquisition

1. Activate the Model: RTModel023A.mdl

2. Connect Matlab RT and run the model.

3. The model will drive the motor and stops after 10 second.

4. Turn OFF the hydraulic power

   Note: This response includes only the valve dynamics and ignores the motor and rpm sensor dynamics.
   Motor is represented only by I-n static characteristics

4

Step 6. Find the Time Constant & Construct normalized T.F.

Guidance to Students

$$\tau = 5 \text{ ms}$$

$$N.T.F = \frac{1}{\tau s + 1} = \frac{1}{0.005s + 1}$$

5

Step 7. Repeat with the RPM sensor

1. Activate the Model: RTModel023B.mdl

2. Connect Matlab RT and run the model.

3. The model will drive the motor and stops after 10 second.

4. Turn OFF the hydraulic power

Note: This response considers the valve, the rpm sensor and the motor dynamics

6

Step 8. Construct normalized T.F.

Guidance to Students

$$\tau = 50 \text{ ms}$$

$$N.T.F = \frac{1}{\tau s + 1} = \frac{1}{0.05s + 1}$$

7

Lab13: Identify Horizontal Cylinder Dynamics

Objective:
This lab is to practice identification of cylinder dynamics

Step 1: Prepare Components:

MSOE#-X	Component Name	QTY	Drawer #
N027	Pressure measurement hose	1	B
N030	4/2 solenoid operated DCV	1	D3
N035	Solenoid cables (M12)	1	D3

0

Step 2: Adjust the PRV to 500 psi:

1. Connect the pressure measurement hose as shown to P1
2. Set the Pp and Qp toggle switches to "Pot" position
3. Adjust pump potentiometers to Pp = max and Qp = max
4. Fully open the PRV
5. Turn ON the hydraulic power
6. Close the PRV gradually until the pressure gauge reads the set value.
7. Turn OFF the hydraulic power

1

Step 3: Electrical Circuit:

1. Connect solenoid cable to DO6

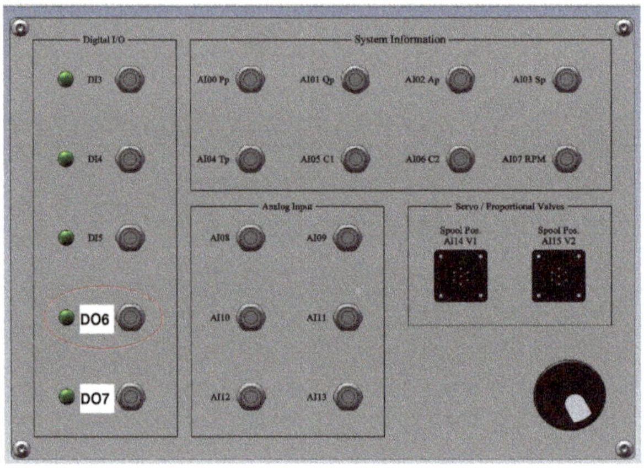

2

Step 4: Hydraulic Circuit:

1. Build the shown hydraulic circuit.
2. Turn ON the hydraulic power.

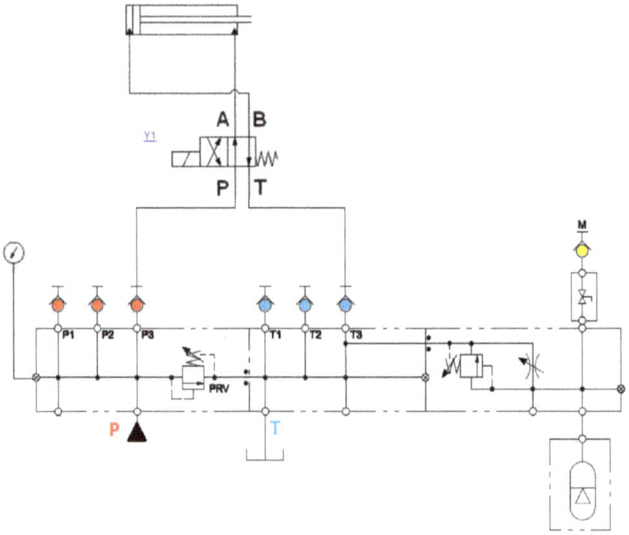

3

Step 5: Data Acquisition:

1. Activate, connect RT and run RTModel017.mdl

2. Model will automatically drive the cylinder and stop after 15 s.

3. Turn OFF the hydraulic power

4

Step 6: Find the time constant & Construct normalized T.F.

- Speed Settling time = 1 s,
- then time const = 1000/5 = 200 ms
- normalized T.F. =

$$N.T.F = \frac{1}{\tau s + 1} = \frac{1}{0.2s + 1}$$

5

Lab14: Proportional Valve Flow Gain Measuring

Objective:
This lab is measure the

Step 1: Prepare Components:

MSOE#-X	Component Name	QTY	Drawer #
N027	Pressure measurement hose	1	B
N029	Proportional DCV + Cable	1	D5
N042 (1-2)/4	Pressure transducer	2	D1
N016-TX-2/4	Flowmeter (Medium Scale)	1	D2
N018	Throttle Valve	1	D2

0

Step 2: Adjust the PRV to 500 psi:

1. Connect the pressure measurement hose as shown to P1
2. Set the Pp and Qp toggle switches to "Pot" position
3. Adjust pump potentiometers to Pp = max and Qp = max
4. Fully open the PRV
5. Turn ON the hydraulic power
6. Close the PRV gradually until the pressure gauge reads the set value.
7. Turn OFF the hydraulic power

1

Step 3: Electrical Circuit:

AI14 V1

1. Connect the proportional valve to the socket AI14V1 and the other side of the cable to the valve.
2. Connect the BNC cable as shown below
3. Put the toggle switch of V1 at "Pot." position
4. Adjust proportional valve potentiometer to "0"
5. Connect pressure sensor (1/4) socket to AIO8
6. Connect pressure sensor (2/4) socket to AIO9

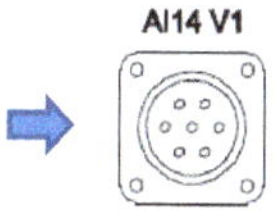

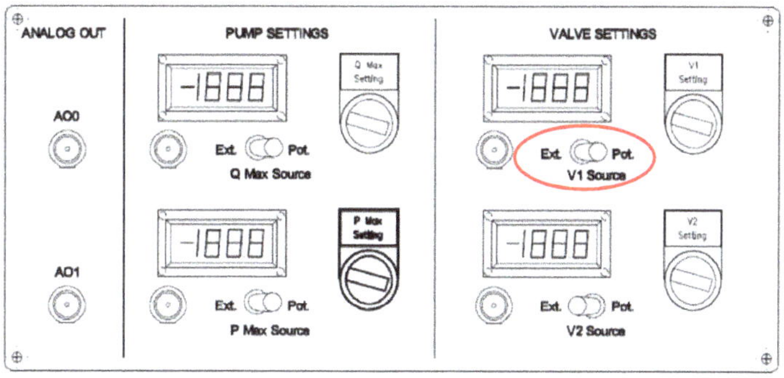

2

Step 4: Hydraulic Circuit:

1. Build the shown hydraulic circuit.
2. Keep the throttle valve fully open
3. Turn ON the hydraulic power.

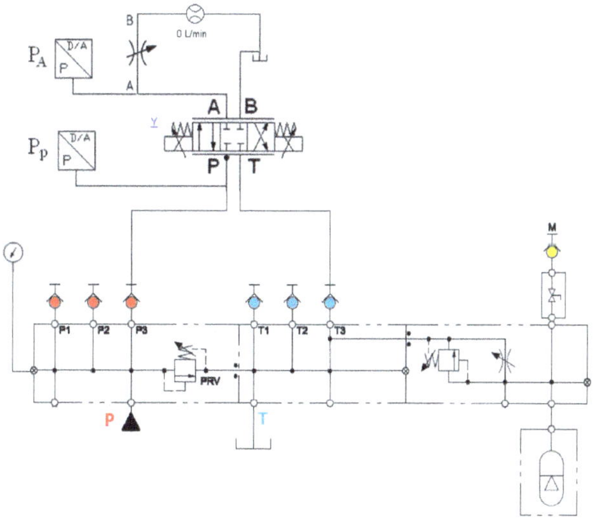

3

Step 5: Data Acquisition:

1. Activate, connect RT and run RTModel018.mdl

2. Adjust the spool position (to the positive side) for the values shown in the table and collect the corresponding flow rate at
   $$DP = (P_P - P_A) = 100 \text{ psi.}$$

(In order to keep DP = 100 psi for each point, open the throttle valve gradually with every increase of the valve position)

Note: Take the readings after being stabilized. Consider the average of the fluctuated readings

3. Turn OFF the power

4. Draw the flow gain best fit curve

4

Constant pressure drop = 100 psi

Q (L/min) vs Spool Position (Volt)

Spool Pos. (Volt)	0	0.5	1	1.5	2	3	3.5	4	5	6	7	8	9	10
Q (L/min)														

5

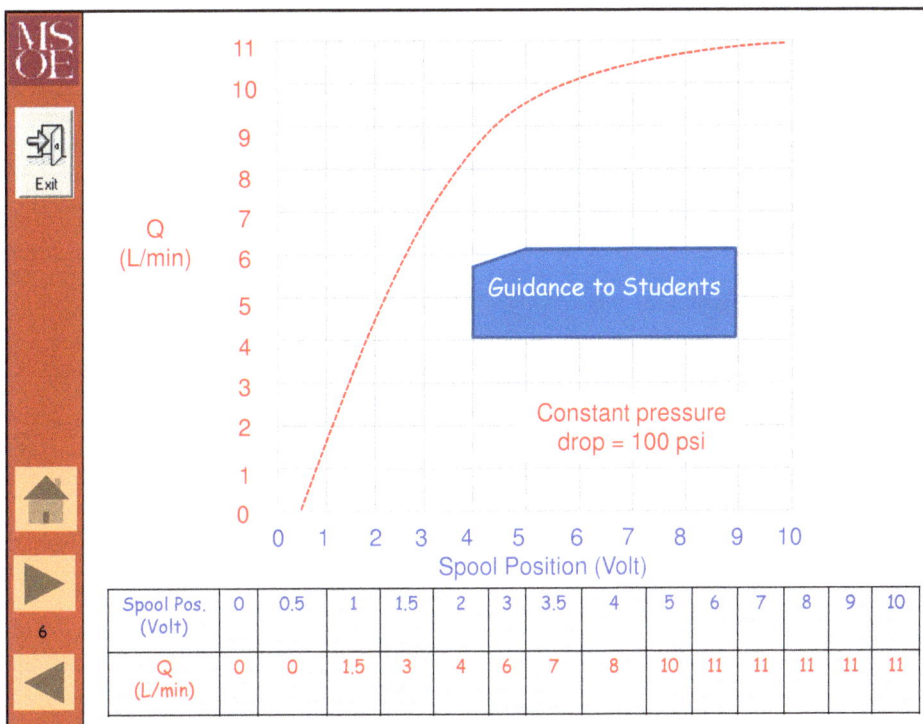

Guidance to Students

Constant pressure
drop = 100 psi

Spool Pos. (Volt)	0	0.5	1	1.5	2	3	3.5	4	5	6	7	8	9	10
Q (L/min)	0	0	1.5	3	4	6	7	8	10	11	11	11	11	11

6

Lab15: Servo Valve Flow Gain Measuring

Objective:
This lab is measure the

Step 1: Prepare Components:

MSOE#-X	Component Name	QTY	Drawer #
N027	Pressure measurement hose	1	B
N033	Servo Valve + Cable	1	D5
N042 (1-2)/4	High Pressure transducer	2	D1
N016-TX-2/4	Flowmeter (Medium Scale)	1	
N018	Throttle Valve	1	D2

0

Step 2: Adjust the PRV to 500 psi:

1. Connect the pressure measurement hose as shown to P1
2. Set the Pp and Qp toggle switches to "Pot" position
3. Adjust pump potentiometers to Pp = max and Qp = max
4. Fully open the PRV
5. Turn ON the hydraulic power
6. Close the PRV gradually until the pressure gauge reads the set value.
7. Turn OFF the hydraulic power

1

Step 3: Electrical Circuit:

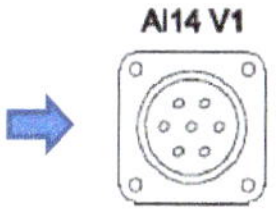

1. Connect the servo valve to the socket AI14V1 and the other side of the cable to the valve.
2. Connect the BNC cable as shown below
3. Put the toggle switch of V1 at "Pot." position
4. Adjust servo valve potentiometer to "0"
5. Connect pressure sensor (1/4) socket to AIO8
6. Connect pressure sensor (2/4) socket to AIO9

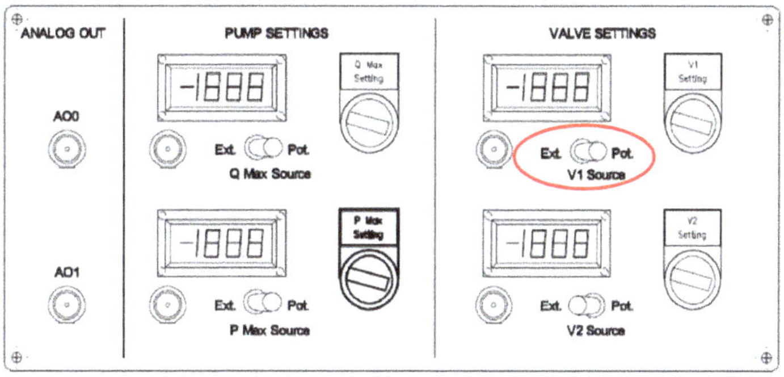

Step 4: Hydraulic Circuit:

1. Build the shown hydraulic circuit.
2. Keep the throttle valve fully closed
3. Turn ON the hydraulic power.

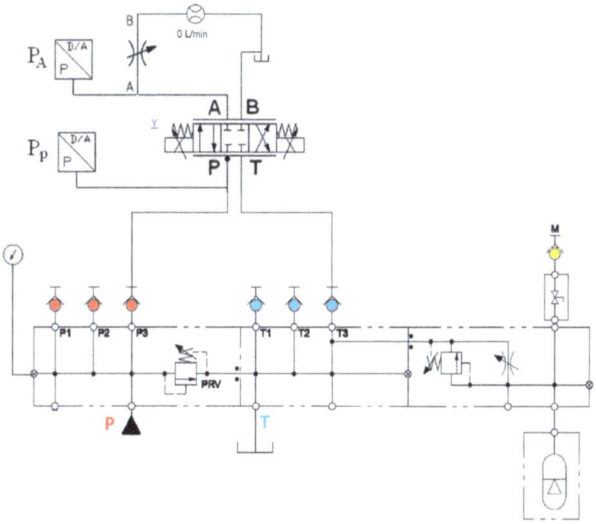

Step 5: Data Acquisition:

1. Activate, connect RT and run RTModel018.mdl

2. Adjust the spool position (to the negative side) for the values shown in the table and collect the corresponding flow rate at DP = 100 psi.

(In order to keep DP = 100 psi for each point, open the throttle valve gradually with every increase of the valve position)

Note: Take the readings after being stabilized. Consider the average of the fluctuated readings

3. Turn OFF the power

4. Draw the flow gain best fit curve

4

Constant pressure drop = 100 psi

Spool Pos. (Volt)	0	1	2	3	4	5	6	7	8	9	10
Q (L/min)											

5

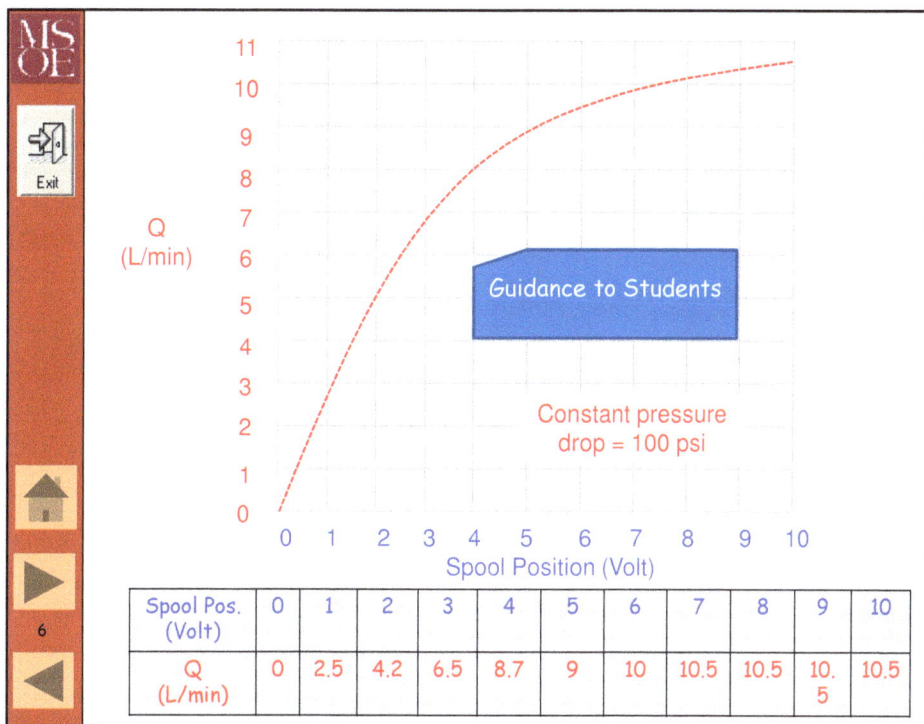

Guidance to Students

Constant pressure
drop = 100 psi

Spool Pos. (Volt)	0	1	2	3	4	5	6	7	8	9	10
Q (L/min)	0	2.5	4.2	6.5	8.7	9	10	10.5	10.5	10.5	10.5

6

Lab16- EH Position Controlled Hydraulic Cylinder Step Response

Objective:

This lab is to practice measuring the step response of a position-controlled hydraulic cylinder.

Step 1: Prepare Components:

MSOE#-X	Component Name	QTY	Drawer #
N027	Pressure measurement hose	1	B
N029	Proportional DCV + Cable	1	D5
N053	BNC Cable	1	D1

0

Step 2: Adjust the PRV to 500 psi:

1. Connect the pressure measurement hose as shown to P1
2. Set the Pp and Qp toggle switches to "Pot" position
3. Adjust pump potentiometers to Pp = max and Qp = max
4. Fully open the PRV
5. Turn ON the hydraulic power
6. Close the PRV gradually until the pressure gauge reads the set value.
7. Turn OFF the hydraulic power

1

Step 3: Electrical Circuit:

1. Connect the proportional valve to the socket AI14V1 and the other side of the cable to the valve.
2. Connect the BNC cable as shown below
3. Put the toggle switch of V1 at "Ext." position

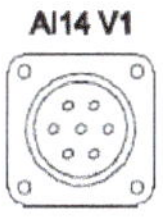

AI14 V1

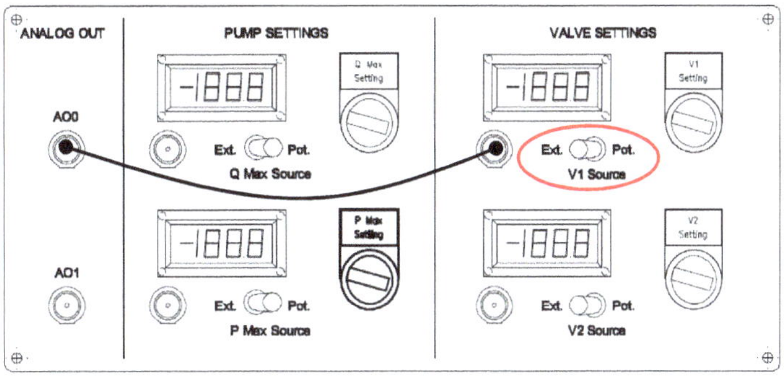

2

Step 4: Hydraulic Circuit:
1. Build the shown hydraulic circuit
2. Connect the pressure sensor to the proportional valve P port
3. Turn ON the hydraulic power.
4. Cylinder must be fully retracted. If not, put V1 toggle switch at "Pot." position, move the potentiometer to retract the cylinder then return back the toggle switch to "Ext." position

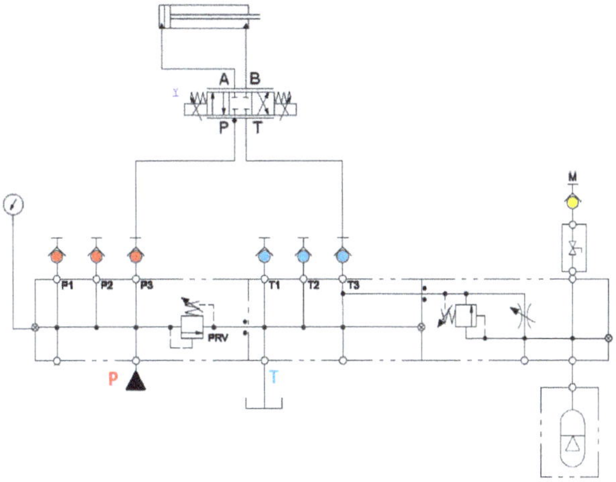

3

Step 5: Data Acquisition, closed loop step response (Dead Band Eliminator)

1. Activate the Model: RTModel019.mdl

2. Click on the function generator and make sure signal is generated from a step input

3. Connect Matlab RT and run the model.

4. The model will drive the cylinder and stops after 10 second.

4

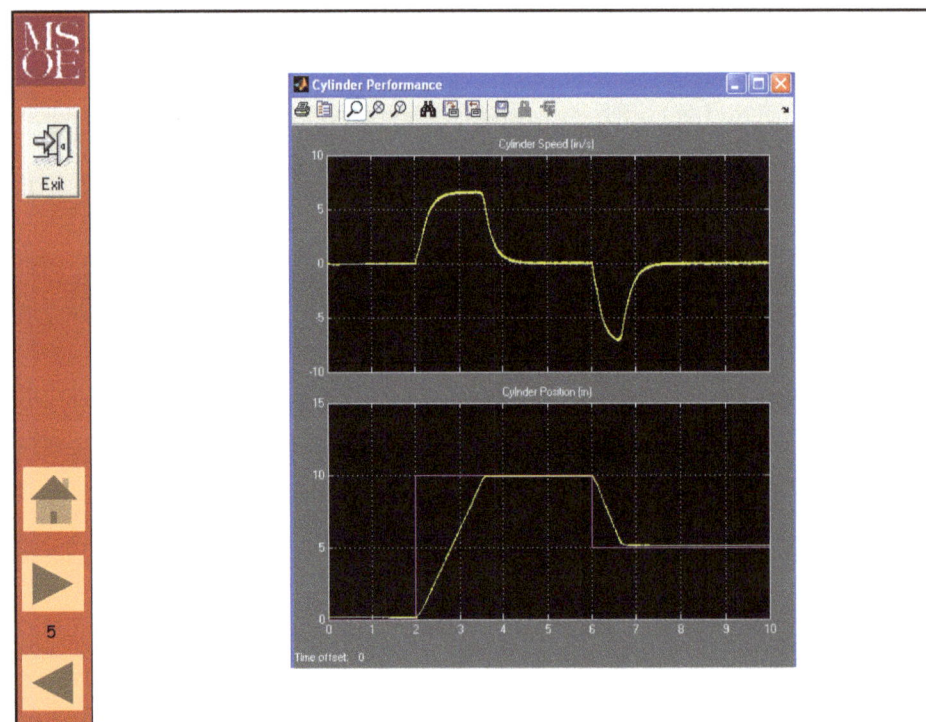

5

Lab17- EH Position Controlled Hydraulic Cylinder Frequency Response

Objective:

This lab is to practice measuring the frequency response of a position-controlled hydraulic cylinder.

Step 1: Prepare Components:

MSOE#-X	Component Name	QTY	Drawer #
N027	Pressure measurement hose	1	B
N029	Proportional DCV + Cable	1	D5
N053	BNC Cable	1	D1

0

Step 2: Adjust the PRV to 500 psi:

1. Connect the pressure measurement hose as shown to P1
2. Set the Pp and Qp toggle switches to "Pot" position
3. Adjust pump potentiometers to Pp = max and Qp = max
4. Fully open the PRV
5. Turn ON the hydraulic power
6. Close the PRV gradually until the pressure gauge reads the set value.
7. Turn OFF the hydraulic power

1

Step 3: Electrical Circuit:

1. Connect the proportional valve to the socket AI14V1 and the other side of the cable to the valve.
2. Connect the BNC cable as shown below
3. Put the toggle switch of V1 at "Ext." position

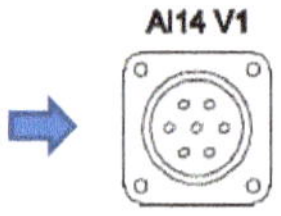

AI14 V1

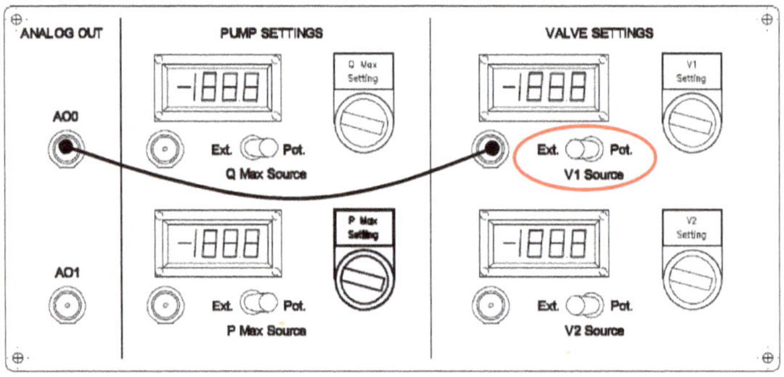

2

Step 4: Hydraulic Circuit:

1. Build the shown hydraulic circuit
2. Connect the pressure sensor to the proportional valve P port
3. Turn ON the hydraulic power.
4. Cylinder must be fully retracted. If not, put V1 toggle switch at "Pot." position, move the potentiometer to retract the cylinder then return back the toggle switch to "Ext." position

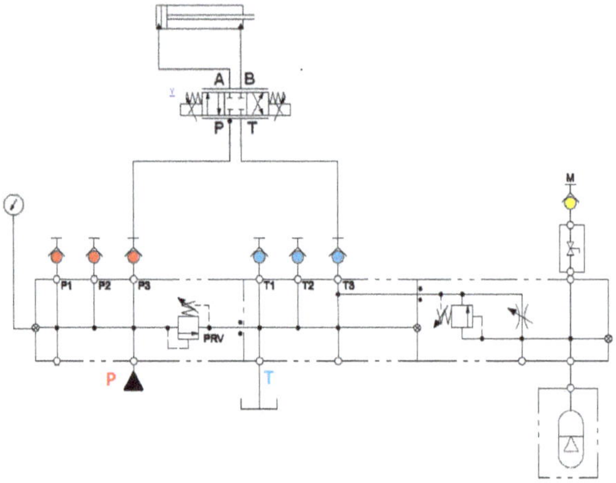

3

Step 5: Data Acquisition, closed loop step response (Dead Band Eliminator)

1. Activate the Model: RTModel019.mdl

2. Click on the function generator and make sure signal is generated from a sinosoidal input

3. Connect Matlab RT and run the model.

4. The model will drive the cylinder and stops after 10 second.

4

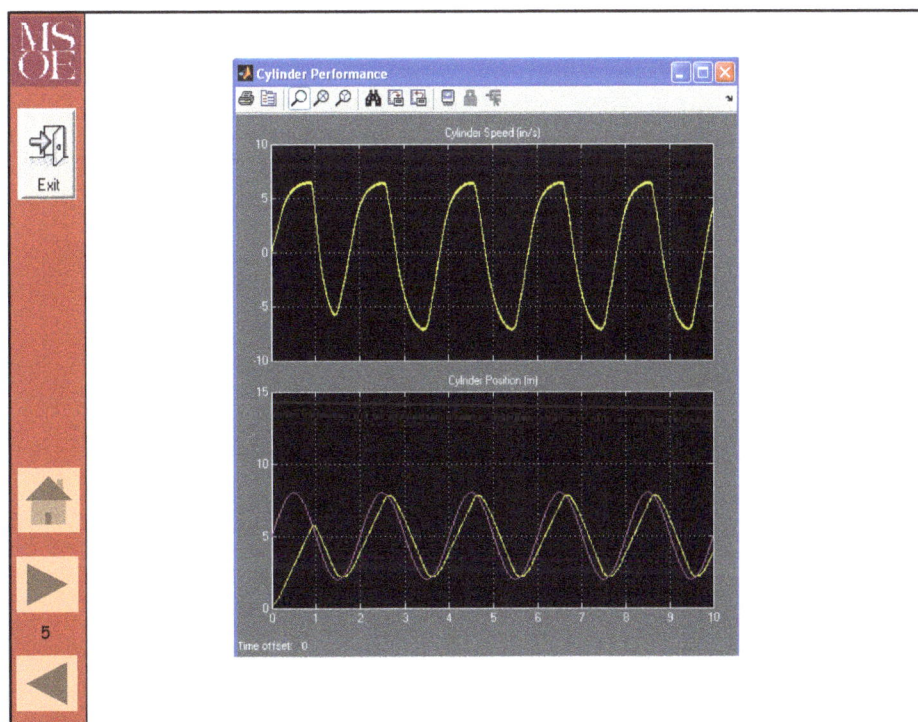

5

Lab18- EH Speed Controlled Hydraulic Motor Step Response

Objective:

This lab is to practice measuring the step response of a speed-controlled hydraulic motor.

Step 1: Prepare Components:

MSOE#-X	Component Name	QTY	Drawer #
N027	Pressure measurement hose	1	B
N033	Servo Valve + Servo Valve Cable	1	D5
N053	BNC Cable	1	D1

0

Step 2: Adjust the PRV to 500 psi:

1. Connect the pressure measurement hose as shown to P1
2. Set the Pp and Qp toggle switches to "Pot" position
3. Adjust pump potentiometers to Pp = max and Qp = max
4. Fully open the PRV
5. Turn ON the hydraulic power
6. Close the PRV gradually until the pressure gauge reads the set value.
7. Turn OFF the hydraulic power

1

Step 3: Electrical Circuit:

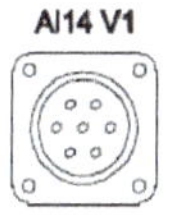

AI14 V1

1. Connect the servo valve to the socket AI14V1 and the other side of the cable to the valve.
2. Connect the BNC cable as shown below
3. Put the toggle switch of V1 at "Ext." position

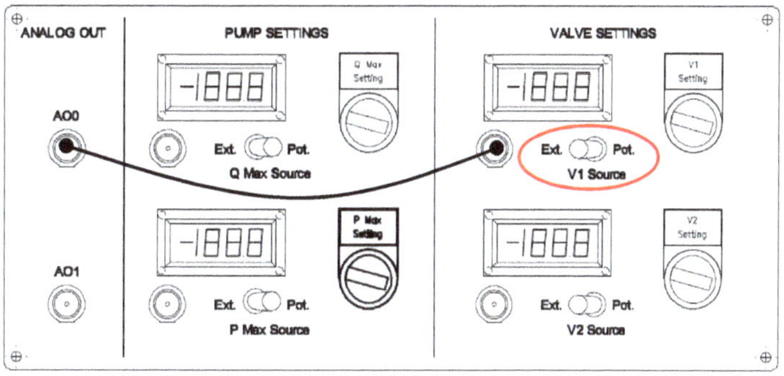

2

Step 4: Hydraulic Circuit:

1- Build the shown hydraulic circuit.
2- Turn ON the hydraulic power.

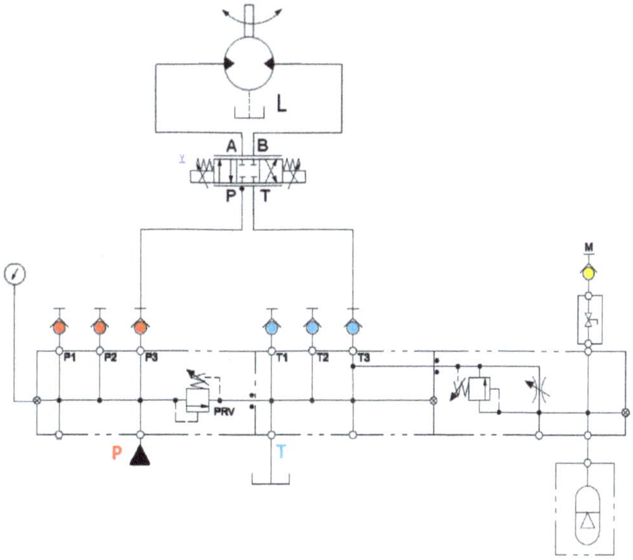

3

Step 5: Data Acquisition, closed loop step response

1. Activate the Model: RTModel020.mdl

2. Click on the function generator and make sure signal is generated from a step input

3. Connect Matlab RT and run the model.

4. The model will drive the motor and stops after 10 second.

- Note: In this RT model with the hardware-in-the-loop, we are not really closing the loop on the rpm sensor. We rely on the calibration I-n to get the rpm.

4

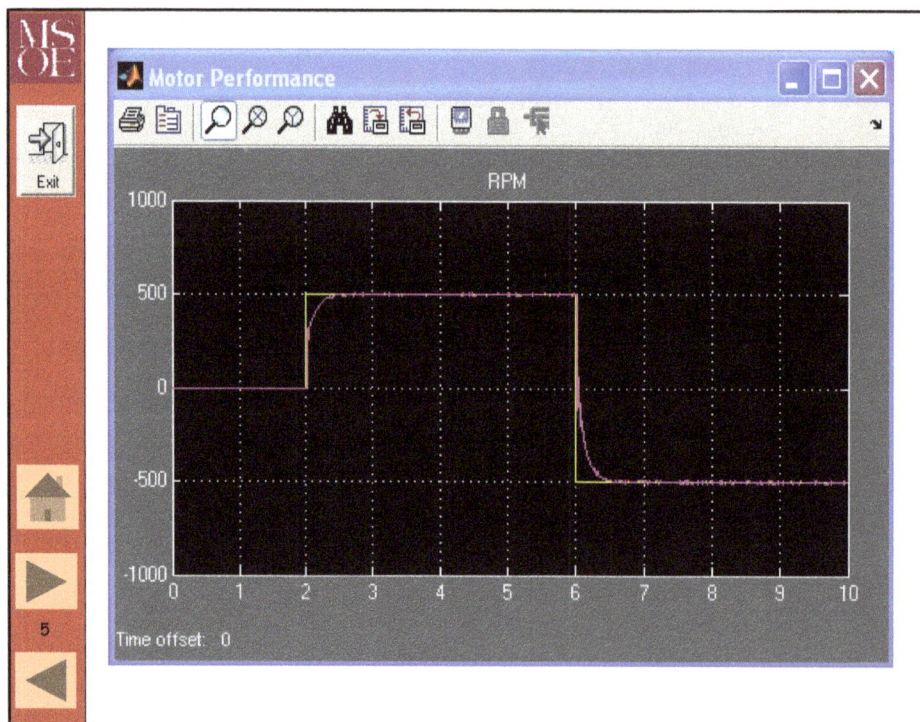

5

Lab19- EH Speed Controlled Hydraulic Motor Frequency Response

Objective:

This lab is to practice measuring the frequency response of a speed-controlled hydraulic motor.

Step 1: Prepare Components:

MSOE#-X	Component Name	QTY	Drawer #
N027	Pressure measurement hose	1	B
N033	Servo Valve + Servo Valve Cable	1	D5
N053	BNC Cable	1	D1

0

Step 2: Adjust the PRV to 500 psi:

1. Connect the pressure measurement hose as shown to P1
2. Set the Pp and Qp toggle switches to "Pot" position
3. Adjust pump potentiometers to Pp = max and Qp = max
4. Fully open the PRV
5. Turn ON the hydraulic power
6. Close the PRV gradually until the pressure gauge reads the set value.
7. Turn OFF the hydraulic power

1

Step 3: Electrical Circuit:

1. Connect the servo valve to the socket AI14V1 and the other side of the cable to the valve.
2. Connect the BNC cable as shown below
3. Put the toggle switch of V1 at "Ext." position

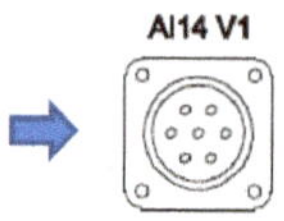

AI14 V1

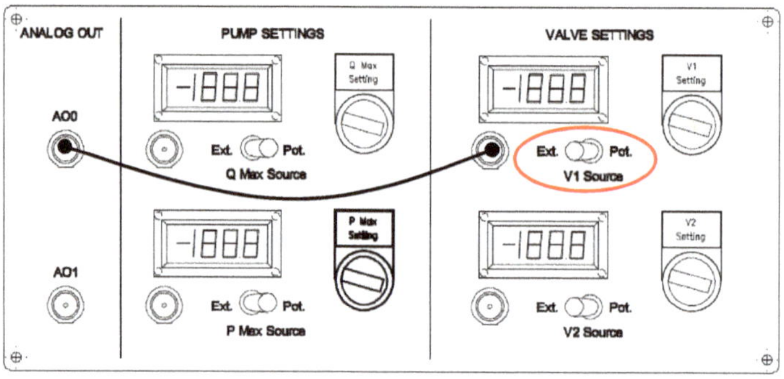

2

Step 4: Hydraulic Circuit:

1- Build the shown hydraulic circuit.
2- Turn ON the hydraulic power.

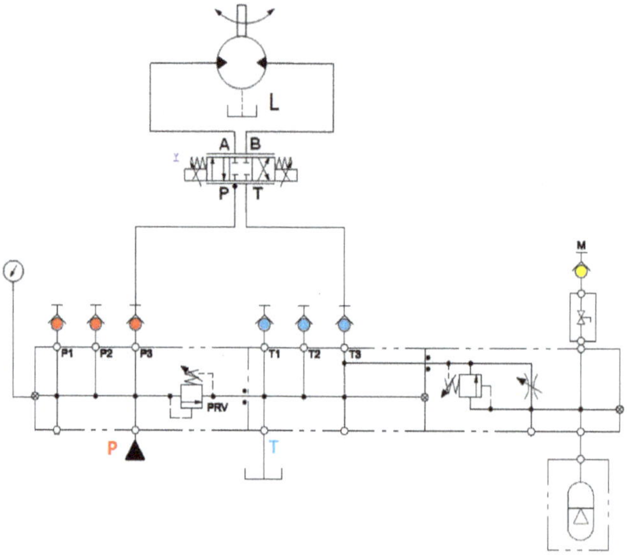

3

Step 5: Data Acquisition, closed loop step response

1. Activate the Model: RTModel020.mdl

2. Click on the function generator and make sure signal is generated from a sinosidal input

3. Connect Matlab RT and run the model.

4. The model will drive the cylinder and stops after 10 second.

- Note: In this RT model with the hardware-in-the-loop, we are not really closing the loop on the rpm sensor. We rely on the calibration I-n to get the rpm.

4

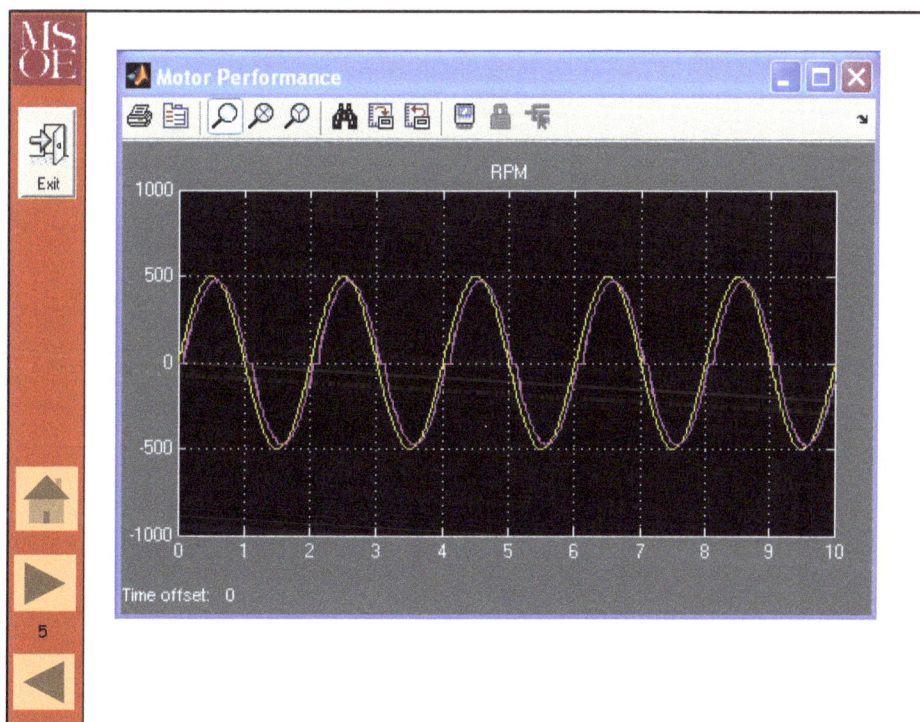

5

www.ingramcontent.com/pod-product-compliance
Lightning Source LLC
Chambersburg PA
CBHW050241220326
41598CB00047B/7471